THE RISE AND FALL OF THE SEAS

The Rise
and Fall
of the Seas

The Story of the Tides

Ruth Brindze

Illustrated with photographs and with diagrams by Felix Cooper

Harcourt, Brace & World, Inc., New York

Library of Congress Catalog Card Number: 64-11491
Printed in the United States of America
First edition

Contents

THE RISE AND FALL OF THE SEAS

1: The Spectacle at the Seashore

From minute to minute — even from second to second — the level of the oceans continually changes. During part of the day, the water becomes deeper. In the following period, it becomes shallower. This alternate rise and fall is called the tide.

In mid-ocean, as well as near land, the water moves vertically, up and down. But the tide's movement is apparent only when there is something against which to measure it. People aboard a ship can neither see the tide's motion nor feel its effect, for a ship rises and falls with the water. But from the shore, you can actually see the water becoming deeper or shallower.

It is not necessary to choose a special day or time for observing the tide. The spectacle can be viewed at any hour of any day. The tide puts on a continuous performance.

On a rising tide, or, as sailors say, when the tide is "making," the water moves steadily farther and farther up the beach. The mark where one wave ends is soon overrun by the waves behind it. And the marks they leave are passed by other waves. The increase in the depth of the water causes it to spread out and submerge more of the land.

The vertical movement of the sea is seen most clearly where

On a rising tide, the mark where one wave ends is overrun by the following waves.

there is a pile of rocks or a pier. As the water rises, rocks or parts of the pier's foundation, which an hour or so before were exposed, are covered by the sea.

In some parts of the world, the rise in the water level is so great that at high tide large ships can steam over areas that at low tide are dry enough to walk on without getting one's feet wet. In the Bay of Fundy, on the east coast of Canada, the water rises more than forty feet in a little more than six hours.

The Bay of Fundy tides are the greatest in the world. But even where the rise is not so extreme, the tide is of vital importance. Ships can enter or leave many ports only because the tide increases the depth of the channels enough to make them navigable. Seafarers made practical use of the tide long before anyone understood why the depth of the oceans is ever changing.

Low tide in the Bay of Fundy. The pier appears to be located on dry land. At high tide, water floods over the end of the pier. The lighthouse appears to be on an island.

Ancient tide table prepared in the thirteenth century shows the time of high tide at London Bridge. The first column gives the moon's age in days. The second and third columns give the hour and minute of high tide after the moon's passage overhead. The ancient table, unlike modern ones, which apply to one particular month, could be used for any month.

Tabula flod at london brigge

Tabula	ho ze	mi minuta
1	3	48
2	4	36
3	5	24
4	6	12
5	7	0
6	7	48
7	8	36
8	9	24
9	10	12
10	11	0
11	11	48
12	12	36
13	1	24
14	2	12
15	3	0
16	3	48
17	4	36
18	5	24
19	6	12
20	7	0
21	7	48
22	8	36
23	9	24
24	10	12
25	11	0
26	11	48
27	12	36
28	1	24
29	2	12
30	3	0

As often happens when true facts are unknown, fantastic stories were told about the tides. According to one, a god living at the bottom of the sea gulped in the water, thus causing a low tide. When he spat out the water, there was a high tide.

This tale, and others equally fanciful, satisfied some people. But those who studied the moon and the sun knew that there is a relationship between the tide and these heavenly bodies. It was noted that soon after a new moon, there was an unusually high tide. About two weeks later, when the full golden moon lit up heaven and earth, the tide again rose unusually high.

Almost two thousand years ago, a Roman scholar named Seneca recorded the fact that the sun, as well as the moon, plays a part in generating higher than normal tides. Seneca observed that exceptionally high tides occur when the sun, moon, and earth are in line.

Another discovery of great practical value was that at each place on the coast the tide rises to its maximum height a certain number of hours after the moon passes overhead. The discovery that the tide follows the moon in a regular pattern made it possible to predict the movement of the tides. One tide table prepared in the thirteenth century predicted the time of high water at London Bridge. The brief, handwritten tide table is on display in the British Museum in London. The table is headed "flod at London brigge." The word "flood" is still used for a rising tide and "ebb" for a falling tide.

Seamen studied the behavior of the tide wherever they went. Explorers were instructed to keep a record of the time of the highest and lowest water along all the shores they sailed and in all the harbors in which they anchored.

Much useful information about the tide was gathered in this way, but why the oceans rise and fall remained a mystery. At one time, it was suggested that the moon acts like a magnet and pulls up the water. Finally, Isaac Newton solved the scientific riddle. In 1687, when Newton published his great work on the law of gravitation, he stated that the tide is caused by the gravitational pull exerted by the moon and the sun.

But why does the water rise higher in one place than in another close by? And why does the tide always rise and fall twice a day in some parts of the world, although in others there is sometimes only a single high and low tide? Only comparatively recently have scientists supplied answers to these questions.

Now we have machines that automatically measure and record the tides. The machines draw pictures showing how many feet and inches the water goes up or down in one hour, in the next hour, and so on. This, and other information, is fed to predicting machines that calculate when high and low water will occur in the future. Official predictions are published annually on how the tide will act along most of the world's coasts.

The rise and fall of the tide cause currents that flow horizontally through the sea. In some places, tidal currents move with such great speed that the water swirls around in large and dangerous whirlpools. In other places where the sea bottom is composed of fine sand or gravel, currents cut deep grooves in the ocean floor. On the west coast of Scotland, there are grooves more than five hundred feet deep.

Tidal currents do not always flow in the same direction. In coastal waters, tidal currents generally flow in one direction for a number of hours, and during the following period, the water runs in the reverse direction. In the open ocean, most tidal currents gradually change the direction of their flow.

Information about the direction in which a tidal current will be running and its speed is of utmost importance to navigators. Current tables are issued annually by the United States government and by many other maritime nations. By consulting a current table, navigators find out when the water will be flowing in the direction they are heading and when it will be flowing against them with a resultant decrease in the ship's over-the-bottom speed.

The tide's ceaseless motion has changed the contour of shorelines. Even rocky cliffs have been worn down. The results can be seen in many places. In the Bay of Fundy area, there are some especially weird examples of the tide's sculpturing. Here cliffs have been cut into separate columns. The bottom of the columns, the parts rubbed by the water each day, form a pedestal-like base above which there is a bulging mass of rock. Trees growing on the top of some of the rocks accent the odd appearance of the columns.

Although the tide wears down land, it can be used to build up land. Such an operation is called "warping," a term that comes from the simple dikes constructed long ago in England to hold the sand and mud carried in by a rising tide. The dikes consisted of rows of wooden stakes between which flexible branches were interlaced or warped. The particles of sand, mud, and small stones brought in by the tide were trapped behind the dikes and gradually piled up until they formed solid land. The wind deposited other material on the new land. Vegetation grew upon it, and eventually the land created by the tide was actually above sea level.

The tide is useful in many other ways. It is an inexhaustible source of power for the operation of machinery. In early times, people living in many towns on the coasts of Europe and America

Animals of the intertidal zone are marvelously adapted for living either under water or in the air.

depended on tide mills to grind corn and other grain. Now engineers have developed a system for using the tide to generate electricity. A tidal hydroelectric plant is being built on the north coast of France and is to begin operation in 1966. It will supply electric current not only to nearby coastal communities but also to towns and cities far inland.

All coastal communities get free garbage and waste disposal service from the tide. Tidal currents carry away all sorts of waste matter and bring back clean water from the open ocean. Were it not for the tide, some shore areas might be uninhabitable because of the accumulation of rubbish in coastal waters.

Shore areas that are covered by water at high tide and exposed to the air at low tide are filled with animal and plant life. The creatures and plants of the intertidal zone are marvelously adapted for living sometimes under water, sometimes in the air.

Did the sea creatures that eventually developed into man become air breathers while inhabiting the zone where they were periodically exposed to the sun and air? It is believed that this is what happened.

Our world would be very different if there were no tides.

2: The Moon Is in Command

Isaac Newton's law of gravitation explains the tides. Newton discovered that everything in the universe, large or small, exerts a pull on everything else.

Not only the moon but also the sun and all other stars are pulling on our planet. And it pulls on them. However, the strength of the pull depends on mass and distance. Of all the heavenly bodies, only the moon and the sun are both massive enough and close enough to our planet to generate tides.

The distance between earth and sun is about 93,000,000 miles and between earth and moon about 239,000 miles. The sun's mass is millions of times greater than the moon's mass, but the moon is so much nearer the earth that it rules the ocean tides. The sun is the junior partner in the tide-generating system. Its tide-producing force is less than half as strong as the moon's.

The moon's pull tends to create a bulge in the ocean directly under it. One of the two daily high tides would occur, according to Newton's law of gravitation, when the moon is most nearly overhead (the time of the moon's upper transit, as astronomers say). Sometimes this is at night when we can see the moon and sometimes during the day when the moon is not usually visible.

The tidal bulge tends to follow the moon as it orbits around the earth.

Newton refers only to the tide-making power of the moon and sun. Geography also plays an important role. Variations in the depth of the oceans, the shape, size, and location of the continents, and many other terrestrial features have an effect on the way the tides act. Newton purposely disregarded geography and the effects of the earth's daily rotation in developing his explanation of the tides. It would be a complete explanation only if our entire planet were covered with deep water, undisturbed by forces other than those created by the heavenly bodies and their movements. Suppose for the present we had that kind of world.

One tidal bulge, creating a high tide, would be in the ocean immediately under the moon. Odd though it may seem, there would be a second bulge in the ocean on the opposite side of the world. There the water would bulge outward, away from the moon, and cause another high tide.

The explanation for the second bulge is that the moon's gravitational pull on the distant ocean is weaker than another force to which the water must also respond. This force, known as centrifugal force, tends to propel things outward. All circular motion creates such a force. The centrifugal force built up by the motion of the earth and the moon overpowers the pull of gravity on the distant ocean, and the water bulges outward.

The sun's gravitational pull also tends to create two tidal bulges, one under it and the other in the ocean on the opposite side of the world. But the bulges created by the sun do not ordinarily have a separate effect. They merely increase or decrease the moon's tide.

In our real world, there are no simple tidal bulges. The movement of the tide is far more complicated. But by thinking of tidal bulges, it is easier to understand the basic causes of the tide.

The tidal bulges represent what scientists call the "tide-producing forces." The real tide is caused by these forces. Their effect, however, is modified by the shape of the land above and below the sea.

The sun's tides occur with clocklike regularity. Since our globe spins around on its axis in twenty-four hours, every place on earth passes at twelve-hour intervals through the sun's tidal bulges. If the sun alone produced our tides, high water would be at about the same time every day, and this would also be true of low water.

But because the moon is in command, the tide rises later each day, for the moon is continually moving eastward on its orbital path around our planet. It takes the earth about fifty additional minutes each day to catch up with and to pass through the moon's

The moon's gravitational pull tends to create a bulge in the ocean under it. Centrifugal force creates a second bulge in the ocean on the opposite side of the world.

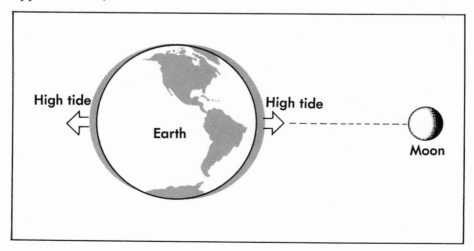

two tidal bulges. Normally, the two daily high tides are spaced about twelve hours and twenty-five minutes apart.

Everyone who spends a vacation at the seashore quickly discovers that the tide's timetable changes daily. In many places the best time to swim is when the tide is high, and on the first morning of your vacation the water may be high at about ten-thirty, which fits into your schedule perfectly. But a few days later high tide will conflict with lunch hour.

Some people's entire life is ruled by the tide. Fishermen know that low tide is the best time to haul their nets, and when low tide is early in the morning, they start work before dawn. Sailors may not finish work until midnight if the tide has only then risen sufficiently so that boats can be brought into port.

In some island communities, even school hours are arranged to fit the tide's schedule. For example, on the North Sea islands called the Halligen, where cattle raising is the major industry, pupils are excused from school whenever it is necessary to round up the herds before the tide submerges the pastures. Sometimes the tide floods the pastures in midmorning, sometimes in the afternoon. School hours are shifted accordingly.

The Halligen are near the German coast, and there is ferry service between the islands and the mainland. Each day, however, the boat arrives and departs at a different time, depending on when the tide makes the channels deep enough to be navigable. At other times, no one can reach the islands or leave them.

Islands with airplane ferry service may also be cut off from the rest of the world for a part of each day. A beach serves as the landing strip for Barra, a Scottish island in the group called the Outer Hebrides. The beach, composed largely of cockleshells, makes an excellent airport at low tide. However, when the tide rises, Barra's airport is under water. Airplanes must take off before the tide rises over the beach, and during the hours that the beach is flooded, no airplane can land.

Anyone who knows the time of high tide on one morning can make a rough guess as to when, at the same place, the water will be high the next morning. It will be about fifty minutes later. But at another beach, a hundred miles away, the time of high tide may be quite different. However, in any one locality the tide generally follows a definite schedule.

Tide tables must be consulted for precise information about the time of high and low water, for the interval between high tide on one morning and the next is not exactly fifty minutes. The height to which the water climbs and the extent to which it recedes also changes from day to day.

During two periods each month, when the moon is new and when the moon is full, the water rises higher and falls lower than at other times. The new moon is not visible in the sky, but we know

The beach at Barra makes an excellent landing strip at low tide.

Radio Times Hulton Picture Library

that moon and sun are in line with each other and with the earth. Both are on the same side of our planet. When the moon is full, it is again lined up with the sun, although the moon is then on one side of our planet and the sun is on the other.

When the moon and sun are in line, they work as a team. The moon's high tide and the sun's high tide come at about the same time. Their low tides also coincide. The result is that there is a greater-than-usual rise and fall in the level of the oceans. High tide is higher than average; low tide is lower. The dates when the moon will be lined up with the sun can be calculated years ahead. Tide tables give the information for each month of the year they cover.

Often when ships are driven ashore by storms, no immediate attempt is made to pull them back into deep water. People who read accounts of such mishaps frequently wonder why, if a rescue tug is on the scene, she does not go to work at once to aid the stranded ship. The strategy is to wait for the next unusually high tide. Then the increase in the depth of the water makes refloating the ship a safer, as well as an easier, operation.

The highest and lowest tides of the month are known as "spring tides," or just "springs." The name comes from the Anglo-Saxon word "springan," meaning to leap. Spring tides occur during all seasons and not merely in the spring.

About a week after a spring tide, the rise and fall of the sea is less than usual. "Neap tides," as they are called, occur when the moon is in its first quarter and in its third quarter. At both times, the moon is at right angles to the sun. Then the two heavenly bodies engage in a tug of war. The lunar high tide comes at the time of the solar low tide. When the sun's tide adds nothing to the moon's tide, the rise in the water level is less than usual. And the low tide

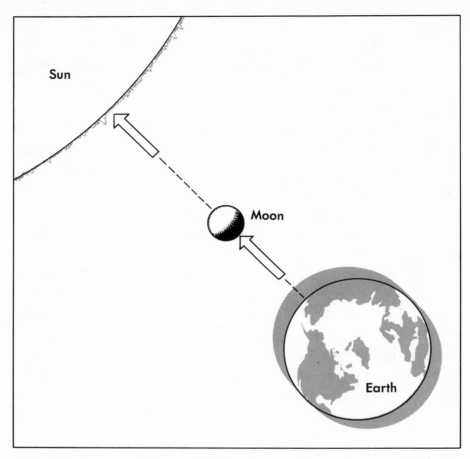

When the moon and sun are in line, they work together as a tidal team. Then the tide rises higher and falls lower than normal.

mark is higher than usual, for when the moon's tide is at its low point, the sun's is at its high point. The result is that the low tide at neaps is not nearly so low as on other days of the month.

Although many calendars tell the dates of the moon's phases — new, first quarter, full, and last quarter— this information alone is not enough to predict the dates of spring or neap tides, for they do not occur on exactly the day of the moon's phase indicated on the calendar. In some places, they occur a day later, in other places, two. The interval is determined from records of the way the tides have acted in the past.

Of course, the tide does not suddenly shoot up to its high spring tide level. The change is gradual. For several days, just before and

When the moon is at right angles to the sun, the two heavenly bodies engage in a tug of war. Then the tide rises and falls less than normal.

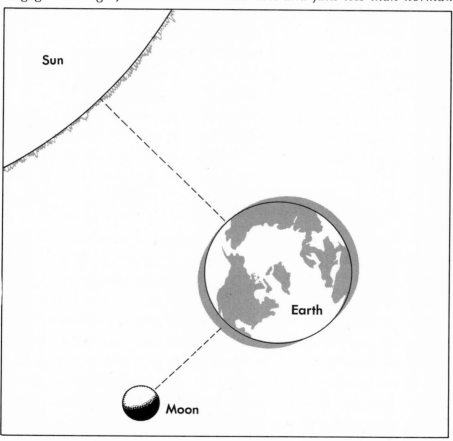

just after the tide reaches its greatest height, it rises higher than usual.

About twice each year the spring tides are exceptionally high. The water rises about 40 per cent higher than on normal days. These extremely high tides occur when the moon is either new or full and at the same time is close to the earth. The moon is not always the same distance from our planet. Each month as the moon orbits around the earth, it at times swoops in closer to us and at other times moves farther out into space. The difference between its nearest and farthest positions is about 30,000 miles.

When the moon is closest, its greater gravitational pull increases the height of the tides. Parts of the shore ordinarily far above the high-water mark may be flooded.

At many beaches the chairs, umbrellas, mats, and other things used during the day are-stacked up on the upper part of the beach when the bathing period is over. On most days the water does not come within many feet of the beach paraphernalia. But on those rare days of extremely high tides, the water may cover the stacks of equipment, and some of it may float away.

The same thing may happen when storms cause higher than normal tides. Warnings about unusually high tides due to storms are included in weather reports. But only a tide table gives advance notice of when the moon's gravitational pull will be at peak strength.

In most parts of the world, there are two high tides and two low tides each day. This is the pattern on the Atlantic Coast of the United States. When the moon is over the equator, the morning and evening high tides at most beaches rise to about the same height. The two low tides are also about equal.

The reason for the equality is that the centers of the moon's tidal bulges are directly opposite each other on the equator. There-

Storm waves at high tide flood over a highway. The army trucks passing the abandoned automobile are going to the rescue of people endangered by the unusually high tide.

fore, during the daily rotation of our planet, a seaport near the equator moves through the center of one tidal bulge and, twelve hours later, through the center of the second bulge. Similarly, places located on any line of latitude pass through the two tidal bulges where they are of about the same height. As a result, the two high tides of the day are very much alike.

This is not the case when the moon is north or south of the equator. Because of the moon's angle to the earth, the center, or highest, portion of one tidal bulge is north of the equator, the other south. Every place on earth during a day passes through a high part of one bulge and a low part of the other. Then the two high tides (and the two low tides) of the day may be quite different.

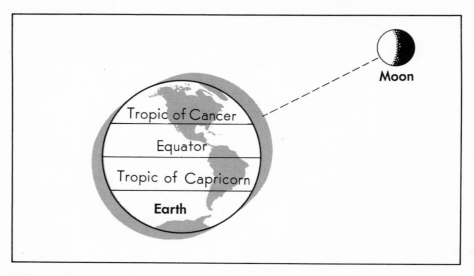

When the moon is over the equator (see diagram on page 20), the centers of its two tidal bulges are opposite each other on the equator. When the moon's position is north or south of the equator, the centers of the two tidal bulges are on different lines of latitude. Then every place on earth passes through a high part of one bulge and a low part of the other bulge.

For instance, at Atlantic City, a popular East Coast beach resort, high tide may occur on one particular day at 9 A.M. If the moon is north of the equator, the morning high tide is caused by the moon's direct pull. The water rises to within a few feet of the "boardwalk," a wide promenade built parallel to the ocean. About twelve hours later, the rotation of the earth brings Atlantic City opposite the moon, and another high tide occurs. But this one will end far below the morning high-tide mark. The evening high tide will leave a wide strip of beach between the boardwalk and the water's edge.

Along some coasts, when the moon is farthest north or south of the equator, the water sometimes rises too little during one high tide to produce any noticeable effect. The level of the water at high tide is about the same as at low tide. In other words, one of the day's high tides vanishes. This happens on parts of the West Coast of the United States. During most of the month there are two high tides and two low tides each day. But on some days there is only a single high and low tide.

Many places situated on the Gulf of Mexico usually have only one high and one low tide a day. The situation is similar at Manila in the Philippine Islands. At Do-Son in Vietnam the tides always rise and fall once daily.

A vast amount of scientific detective work was required before the tide's behavior in different parts of the world could be explained. The movement of the moon north and south of the equa-

The shaded portions represent the oceans' basins in which the tide rises and falls twice each day. In the unshaded areas, there is little tidal activity. The Roman numerals indicate the hour of high tide based on the time of the moon's passage over Greenwich, England.

From the Manual of Tides by R. A. Harris

tor alone cannot create the once-a-day high and low tides on some coasts.

The difference in the height to which the water rises in comparatively nearby places also seems puzzling. The Panama Canal is only about thirty miles long; yet at the Atlantic end, the tide rises less than one foot and at the Pacific end from twelve to sixteen feet. The east coast of the Korean Peninsula, which faces the Japan Sea, has very small tides. The average rise is less than one foot. Korea's west coast is on the Yellow Sea and has a tidal rise averaging more than twenty feet.

Scientists now believe that the once-a-day high tides and the great variations in the heights to which the tide climbs occur because the oceans are divided into natural basins, each of which responds differently to the tide-producing forces of the moon and sun. In each of the natural basins, the water sloshes back and forth at a speed determined by the length and depth of the basin.

If you put water into a tank and raise one end, the water will flow to the low end. If you then raise that end, the water will run down to the opposite end. By lifting first one end of the tank and then the other, you will create what scientists call a "stationary wave."

With the water level shown from A-B, high tide is between A-N. Low tide is between N-B. With the water level shown from C-D, high tide is between C-N. Low tide is between N-D. The water level at N, the nodal line, remains the same whether it is high tide at the left or right end of the tank.

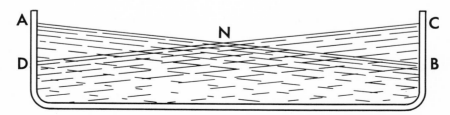

In the oceans' natural basins, water moves in stationary waves. The tide-making forces of the moon and sun keep the waves moving.

When a stationary wave is up at one end of its basin, high tide occurs there. At the other end, it is low tide. At places near either end of a natural basin, the tidal rise and fall is greatest. The effect is exactly the same as when a tank is rocked and the water piles up at one end.

The Pacific entrance of the Panama Canal is at the end of a stationary wave system, and the water piles up there. At the Atlantic entrance, the small rise and fall of the water is typical of the basin that contains the Gulf of Mexico and the Caribbean Sea. Because of the dimensions of this basin, its stationary wave does not always move back and forth twice a day. Frequently, in the Gulf and the Caribbean Sea, there is only one daily high and low tide.

If you experiment with a tank, you will observe that no matter which end is raised, the water level at the center always remains about the same. The same thing happens in a natural basin. The result is that places located near the center, or nodal, line of a natural basin have very small tides.

The island of Tahiti is near the center line of a stationary wave system sparked by the moon's tide-producing power. Therefore, the tides caused by the moon are small. However, the basin in which Tahiti lies does respond to the sun's pull. Since the sun controls the rise and fall of the sea on the shores of the South Pacific island, high tide and low tide occur at about the same time every day. High tide is about noon and midnight and low tide about 6 A.M. and 6 P.M.

The great tides in the Bay of Fundy are caused partly by the character of its stationary wave and partly by visible geographical conditions. The bay is shaped like a funnel. At its mouth, it is eighty-seven miles wide and about two hundred and eighty feet deep. From there on, it becomes progressively narrower and shallower. In the upper part of the bay, where its north and south shores are only thirty miles apart, a peninsula divides the bay into two inlets. In these, the channels are even narrower, and the water is forced to rise in order to move ahead.

Not only the formation of the land but also the movement of the stationary wave in the Bay of Fundy contribute to the exceptionally high tides at the head of the bay. The mouth of the bay is near the center line of a natural basin, and there the tidal rise is comparatively small, averaging only about nine feet. The head of the bay is at an end of the basin, where the stationary wave reaches its maximum height. The tidal rise at the head of the bay averages more than forty feet. The rise is very rapid, sometimes more than eleven feet in an hour. Along the Atlantic Coast, there are many places where the water level does not increase as much in six hours.

The effect of the stationary wave in the Bay of Fundy and the existence of stationary waves in the world's oceans have been proved mathematically. Mathematics is important in all tidal work. Predictions of how much the water will rise or fall on any future date are based on elaborate calculations. Provided that certain facts are known, the action of the tides in the future can be predicted with great accuracy.

3: How Tides Are Predicted

The historic date known as D Day, when the Allies invaded France during World War II, was chosen because the tide was right for landing an army on the beaches of Normandy. On June 6, 1944 — D Day — there was a spring tide.

For months, General Dwight D. Eisenhower, Supreme Allied Commander, and his staff of military experts had been planning the invasion. They had studied reports on the German defense system, on the character of the waves and currents in the English Channel, and on the behavior of the tide on the north coast of France. They knew that the beaches were heavily mined and that many booby traps had been set on them.

On Omaha Beach and on other landing sites, the water rises over twenty-two feet when there are spring tides. The unusually high rise is followed by a greater-than-average fall in the water level. At low tide, nearly half a mile of sand, mud, and rock is exposed, which at high tide is covered by water. The strategy decided upon by General Eisenhower and his staff was to begin the invasion when the greatest expanse of beach was bare of water so that the mines and traps could be seen and destroyed.

According to the tide table, there were three days in the first

The traps on the beaches of Normandy as seen at low tide.

week of June, the fifth, sixth, and seventh, when tidal conditions were favorable for the invasion. The next spring tides would not occur for two weeks.

June 5 was General Eisenhower's first choice for D Day. In accordance with his orders, some troopships had left port and were steaming toward the French coast when a heavy gale began to roar over the English Channel. There was no choice but to recall the ships and to postpone the invasion. It then appeared that the storm would continue for many days. Fortunately, the weather picture changed. Immediately after receiving a good forecast for June 6, General Eisenhower announced, at a dramatic meeting of his staff, "We'll go."

There was a spring tide during the first week of June because there was a full moon. The moon rose late, which also was of strategic advantage. Airplanes carrying paratroops could approach their objectives in darkness, but there would be moonlight after the airborne troops parachuted to the ground.

Low tide was early in the morning and late in the afternoon. During the morning low tide, landing craft brought in the first assault troops and their equipment to the beaches. In the afternoon, reinforcements were put ashore.

The basic tidal information on which General Eisenhower relied was not prepared especially for the invasion. Much of the data had been calculated by the British Admiralty as part of its routine work. During peace as well as during war, every important maritime nation compiles tidal predictions.

Each year, the United States Coast and Geodetic Survey issues four paper-covered books containing information about the rise and fall of the tide at about six thousand places in various parts of the world. Every ship carries official tide tables. When a captain is approaching a port, he first checks a nautical map to find the depth of the water in the entrance channels. Usually, the depths are based on the level at low tide. Even though at low tide the water is not deep enough for his ship, a captain can determine, by referring to a tide table, when the tide will increase the water level enough so that he can enter the port safely.

For example, suppose the captain of a large fishing boat decides that the best place to sell his catch is at Jonesport, Maine. The deep-sea fishing boat, a sturdy craft designed to operate far offshore, cannot be navigated in water less than fifteen feet deep.

According to the nautical map, the water in the channel leading to Jonesport is, in places, only ten feet deep at the time of low tide. The captain then looks up Jonesport in the tide table and finds the following information. High tide at Jonesport is twenty-three minutes earlier than at Portland, Maine, and the water rises 2.6 feet higher than at Portland.

Portland is the reference station for Jonesport and other places on the coasts of Maine, New Hampshire, and Massachusetts where the tidal pattern is similar. From the predictions of high and low water at a reference station, it is easy to figure the state of the tide at related harbors. If the reference station system were not used and full details were supplied for every harbor for every day of the year, tide tables would be as bulky as big dictionaries.

The captain turns to a page headed Portland and finds that the evening's high tide is predicted for 9:25 P.M. At Jonesport, where the tide is twenty-three minutes earlier than at Portland, high tide will be at 9:02 P.M. (9:25 minus 23 minutes). In another column on the Portland page, the captain sees that the tide will rise 8.2 feet. To this figure, he adds 2.6 feet to obtain the high-tide level at Jonesport. At high tide, the depth of the channel will be increased by 10.8 feet.

The figures on the nautical map give the depth of the channel to Jonesport at low tide.
          U. S. Coast and Geodetic Survey Chart 304

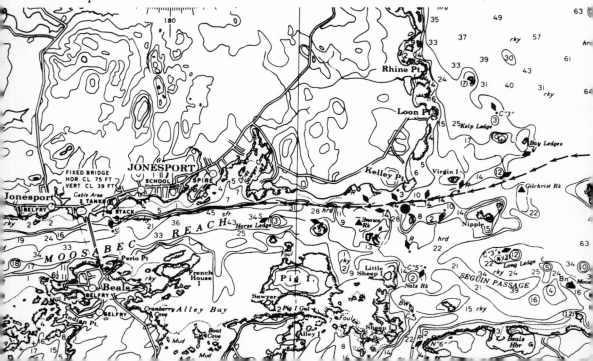

But at what time will the rising tide increase the depth of the Jonesport channel by five feet bringing it up to the fifteen feet required by the fishing boat? Since the tide does not rise or fall at a uniform rate, information is included in all tide tables for calculating the height of the tide at any time. By applying the information, the captain of the fishing boat finds that, according to the predictions, the water in the Jonesport channel will rise five feet by about seven o'clock.

From his long sea-going experience, the captain believed that before that time the water in the channel would be deep enough for his boat, since a strong wind blowing toward the shore was driving in the water. When there is a strong onshore wind, the tide usually begins to rise earlier than the predicted time, and it may rise to a greater height than that listed in the tide table.

A strong wind blowing from the land toward the sea also changes the tidal schedule. An offshore wind may cause the tide to fall longer than the predicted time. Navigators always must consider actual conditions when using the information in tide tables.

No one nation gathers all the facts for its tide tables. The United States Coast and Geodetic Survey collects the data for the coastal areas of this country and for a number of islands. Information for other parts of the world is obtained through an exchange arrangement with other nations. For instance, some of the predictions in the book of tide tables entitled "East Coast of North and South America including Greenland" are based on data supplied by Canada and Argentina.

The first official tide table was issued by the British Admiralty in 1833. Six years later, France began to publish annual tide tables. The United States started to supply tidal information compiled by

PORTLAND, MAINE, 1964

Times and Heights of High and Low Waters

OCTOBER

Day	Time (h.m.)	Ht. (ft.)	Day	Time (h.m.)	Ht. (ft.)
T 1	0053 / 0706 / 1313 / 1928	-0.2 / 8.4 / 0.6 / 9.4	F16	0054 / 0707 / 1311 / 1919	0.9 / 7.5 / 1.6 / 8.1
F 2	0156 / 0806 / 1416 / 2030	-0.4 / 8.7 / 0.2 / 9.7	S17	0142 / 0755 / 1401 / 2008	0.7 / 7.9 / 1.1 / 8.5
S 3	0252 / 0902 / 1512 / 2124	-0.6 / 9.1 / -0.2 / 9.8	S18	0226 / 0837 / 1447 / 2054	0.3 / 8.4 / 0.5 / 8.9
S 4	0342 / 0950 / 1603 / 2213	-0.7 / 9.4 / -0.5 / 9.8	M19	0307 / 0916 / 1529 / 2137	0.0 / 9.0 / -0.1 / 9.3
M 5	0427 / 1034 / 1649 / 2258	-0.7 / 9.6 / -0.7 / 9.7	T20	0348 / 0955 / 1612 / 2220	-0.3 / 9.5 / -0.7 / 9.7
T 6	0508 / 1115 / 1732 / 2339	-0.5 / 9.7 / -0.8 / 9.4	W21	0427 / 1035 / 1655 / 2303	-0.6 / 10.0 / -1.2 / 9.8
W 7	0547 / 1153 / 1813	-0.3 / 9.7 / -0.6	T22	0509 / 1118 / 1740 / 2348	-0.7 / 10.4 / -1.5 / 9.8
T 8	0019 / 0624 / 1232 / 1854	9.1 / 0.1 / 9.5 / -0.3	F23	0553 / 1202 / 1828	-0.7 / 10.6 / -1.6
F 9	0100 / 0702 / 1311 / 1936	8.6 / 0.4 / 9.2 / 0.0	S24	0037 / 0640 / 1250 / 1919	9.6 / -0.5 / 10.5 / -1.5
S10	0142 / 0742 / 1352 / 2020	8.2 / 0.9 / 8.9 / 0.3	S25	0128 / 0731 / 1342 / 2014	9.3 / -0.2 / 10.3 / -1.2
S11	0226 / 0825 / 1436 / 2108	7.7 / 1.3 / 8.5 / 0.7	M26	0226 / 0828 / 1441 / 2115	8.9 / 0.2 / 9.9 / -0.8
M12	0317 / 0915 / 1526 / 2201	7.4 / 1.7 / 8.2 / 1.0	T27	0330 / 0932 / 1546 / 2223	8.5 / 0.6 / 9.5 / -0.4
T13	0414 / 1011 / 1622 / 2259	7.1 / 2.0 / 7.9 / 1.1	W28	0438 / 1043 / 1658 / 2332	8.3 / 0.8 / 9.2 / -0.2
W14	0514 / 1113 / 1723 / 2358	7.0 / 2.1 / 7.8 / 1.1	T29	0548 / 1157 / 1810	8.3 / 0.7 / 9.0
T15	0614 / 1215 / 1823	7.1 / 1.9 / 7.9	F30	0039 / 0655 / 1305 / 1919	-0.2 / 8.5 / 0.5 / 9.1
			S31	0140 / 0754 / 1407 / 2019	-0.2 / 8.9 / 0.1 / 9.2

NOVEMBER

Day	Time (h.m.)	Ht. (ft.)	Day	Time (h.m.)	Ht. (ft.)
S 1	0234 / 0846 / 1502 / 2111	-0.3 / 9.1 / -0.2 / 9.2	M16	0138 / 0752 / 1408 / 2016	0.4 / 8.7 / 0.2 / 8.7
M 2	0321 / 0931 / 1550 / 2158	-0.3 / 9.4 / -0.5 / 9.1	T17	0224 / 0837 / 1457 / 2104	0.0 / 9.4 / -0.5 / 9.1
T 3	0404 / 1013 / 1634 / 2239	-0.2 / 9.6 / -0.6 / 9.0	W18	0310 / 0921 / 1543 / 2152	-0.3 / 10.0 / -1.2 / 9.4
W 4	0443 / 1051 / 1714 / 2319	0.0 / 9.6 / -0.6 / 8.8	T19	0356 / 1006 / 1632 / 2240	-0.6 / 10.5 / -1.6 / 9.6
T 5	0520 / 1127 / 1752 / 2357	0.2 / 9.6 / -0.5 / 8.5	F20	0442 / 1053 / 1721 / 2330	-0.8 / 10.9 / -1.9 / 9.7
F 6	0555 / 1202 / 1829	0.5 / 9.4 / -0.4	S21	0530 / 1141 / 1812	-0.8 / 11.0 / -2.0
S 7	0034 / 0631 / 1238 / 1907	8.2 / 0.7 / 9.2 / -0.1	S22	0021 / 0622 / 1233 / 1905	9.6 / -0.7 / 10.9 / -1.8
S 8	0114 / 0708 / 1316 / 1949	7.9 / 1.0 / 8.9 / 0.1	M23	0116 / 0716 / 1326 / 2003	9.3 / -0.4 / 10.8 / -1.5
M 9	0156 / 0750 / 1358 / 2032	7.6 / 1.3 / 8.6 / 0.4	T24	0214 / 0815 / 1428 / 2103	9.0 / 0.0 / 10.1 / -1.1
T10	0243 / 0837 / 1445 / 2119	7.4 / 1.6 / 8.3 / 0.7	W25	0317 / 0920 / 1534 / 2207	8.7 / 0.3 / 9.5 / -0.6
W11	0334 / 0929 / 1537 / 2211	7.2 / 1.9 / 8.0 / 0.9	T26	0423 / 1031 / 1644 / 2313	8.5 / 0.6 / 9.1 / -0.3
T12	0428 / 1026 / 1634 / 2304	7.2 / 1.9 / 7.8 / 0.9	F27	0531 / 1143 / 1754	8.5 / 0.6 / 8.8
F13	0523 / 1127 / 1733 / 2358	7.3 / 1.8 / 7.8 / 0.8	S28	0016 / 0634 / 1250 / 1901	-0.1 / 8.7 / 0.4 / 8.6
S14	0616 / 1225 / 1831	7.6 / 1.4 / 8.0	S29	0116 / 0732 / 1351 / 2001	0.1 / 9.1 / 0.1 / 8.6
S15	0050 / 0705 / 1318 / 1926	0.6 / 8.1 / 0.9 / 8.3	M30	0209 / 0823 / 1446 / 2053	0.1 / 9.1 / -0.1 / 8.5

DECEMBER

Day	Time (h.m.)	Ht. (ft.)	Day	Time (h.m.)	Ht. (ft.)
T 1	0257 / 0909 / 1533 / 2140	0.2 / 9.3 / -0.3 / 8.4	W16	0145 / 0800 / 1425 / 2035	0.1 / 9.6 / -0.7 / 8.8
W 2	0340 / 0950 / 1616 / 2222	0.3 / 9.4 / -0.4 / 8.3	T17	0236 / 0851 / 1520 / 2128	-0.2 / 10.2 / -1.3 / 9.1
T 3	0419 / 1027 / 1656 / 2301	0.4 / 9.5 / -0.5 / 8.2	F18	0328 / 0942 / 1613 / 2221	-0.5 / 10.7 / -1.8 / 9.4
F 4	0456 / 1103 / 1733 / 2338	0.6 / 9.4 / -0.4 / 8.1	S19	0420 / 1033 / 1705 / 2316	-0.8 / 11.1 / -2.1 / 9.5
S 5	0530 / 1138 / 1808	0.7 / 9.3 / -0.3	S20	0513 / 1126 / 1758	-0.9 / 11.2 / -2.2
S 6	0014 / 0605 / 1212 / 1845	7.9 / 0.9 / 9.2 / -0.2	M21	0006 / 0606 / 1220 / 1852	9.6 / -0.8 / 11.1 / -2.0
M 7	0051 / 0643 / 1249 / 1922	7.8 / 1.1 / 9.0 / -0.1	T22	0102 / 0703 / 1315 / 1948	9.4 / -0.6 / 10.7 / -1.7
T 8	0130 / 0723 / 1328 / 2001	7.7 / 1.2 / 8.7 / 0.1	W23	0158 / 0802 / 1414 / 2045	9.2 / -0.3 / 10.2 / -1.2
W 9	0210 / 0806 / 1412 / 2043	7.6 / 1.4 / 8.4 / 0.3	T24	0258 / 0904 / 1516 / 2143	9.0 / 0.0 / 9.6 / -0.7
T10	0256 / 0853 / 1459 / 2128	7.5 / 1.5 / 8.2 / 0.5	F25	0400 / 1010 / 1621 / 2244	8.8 / 0.2 / 9.0 / -0.2
F11	0343 / 0945 / 1550 / 2216	7.5 / 1.5 / 8.0 / 0.6	S26	0503 / 1119 / 1728 / 2345	8.7 / 0.4 / 8.5 / 0.2
S12	0433 / 1040 / 1646 / 2306	7.7 / 1.3 / 7.9 / 0.6	S27	0604 / 1225 / 1835	8.7 / 0.4 / 8.1
S13	0525 / 1138 / 1744 / 2359	8.0 / 1.0 / 8.0 / 0.6	M28	0045 / 0702 / 1327 / 1936	0.4 / 8.7 / 0.3 / 7.9
M14	0617 / 1236 / 1843	8.4 / 0.6 / 8.1	T29	0139 / 0756 / 1423 / 2030	0.6 / 8.8 / 0.2 / 7.8
T15	0053 / 0709 / 1331 / 1940	0.4 / 9.0 / 0.0 / 8.4	W30	0229 / 0843 / 1512 / 2119	0.7 / 9.0 / 0.0 / 7.8
			T31	0315 / 0926 / 1556 / 2203	0.8 / 9.1 / -0.1 / 7.8

Time meridian 75° W. 0000 is midnight. 1200 is noon.

Heights are reckoned from the datum of soundings on charts of the locality which is mean low water.

A page from the Tide Tables *containing predictions for Portland, Maine. The information includes the times and heights of high and low waters and the number of feet that the water level will rise and fall.* Tide Tables *and* Current Tables *are issued annually by the U. S. Coast and Geodetic Survey and may be ordered from its main office in Washington, D. C., or purchased from authorized sales agents located in coastal communities.*

the Coast and Geodetic Survey in 1853. Unlike modern tide tables, the early ones did not predict the hour and minute of high and low water. They merely provided facts from which seamen could make their own predictions. Only after the rise and fall of the tide had been accurately measured in many places did it become possible to make the kind of tide tables we use today.

The simplest instrument used to measure the up-and-down movement of the oceans is a long ruler called a tide staff. It is marked off into feet, and each foot is subdivided into tenths. A tide staff is fastened to a piling or to some other vertical support, and its exact position is carefully checked and recorded. In order to obtain useful information from a tide staff, it must be read at frequent intervals. The official directions state that the readings should be recorded every half hour or hour except near the times of high and low water. Then it is recommended that the water level be noted at fifteen-minute intervals.

A tide staff installed near a cliff. The staff, marked off into feet and tenths of feet, shows the changing level of the water.

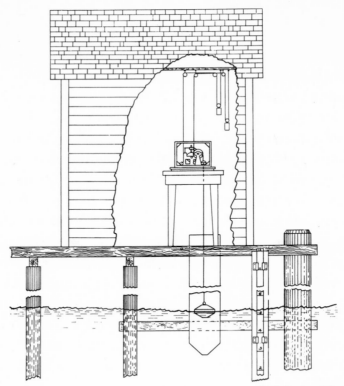

Sketch of an automatic tide gauge showing the float suspended in the water below the recording instrument. As the float rises and falls with the tide, it moves a pencil in the recording instrument. The pencil draws a continuous picture, or graph, of the tide's movement.

Imagine how difficult it would be to make such observations day after day. Only if there were a team of observers so that one could sleep while the other was on duty could the job be handled.

The practical solution is to install at tide-recording stations both a tide staff and an automatic gauge. The gauge draws a continuous picture, or graph, of the rise and fall of the tide. The self-recording gauge generally used is operated by clocks, which run for eight days before requiring winding. However, at many tide-recording stations a daily check is made, and as part of the regular routine, the clocks of the automatic tide gauge are wound.

When an observer visits a tide-recording station, he reads the water level shown on the tide staff and writes it on the drawing that the machine is making. The tide staff provides the basic measurements for interpreting the record made by the automatic gauge.

At the end of each month, the drawings are removed from the machine, they are analyzed, and the results are tabulated. There are many forms to be filled out. On one, the height of the tide at each hour during the month is listed. On another form, the hour and minute of each high and low tide are written in one column and, in the next, the time of the moon's passage over the station. The interval, called the lunitidal interval, must be known in order to make tidal predictions.

Scientifically accurate predictions cannot be made from records of the way the tides have acted in any one year. In some years, for instance, there may be more than the average number of ocean storms that cause the tides to rise higher than usual. Also, the relationships among earth, moon, and sun are not constant from year to year. Therefore, only by measuring the tides for a number of years can an accurate picture be obtained for any one particular area. Tidal scientists prefer to use records for nineteen years.

When the moon will be new or full next year or in the year 2000 is easily determined. The same is true for the dates when the moon's position will be nearest or farthest from the earth. Likewise, the position of the earth as it travels around the sun can be charted years ahead of time. But although all the facts required are available, the mathematical computations for predicting the tides are so complicated that a large staff of skillful mathematicians would have to work day and night to figure out the answers.

About a hundred years ago, Sir William Thomson, a noted British physicist, invented a machine to do the work. The machine drew curves showing the predicted rise and fall of the tide. It was then necessary to translate the curves into the figures needed for tide tables.

The first American tide-predicting machine was designed by William Ferrel, a scientist employed by the Coast and Geodetic Survey. Ferrel's machine was completed in 1882. It did all the computations and gave the time and height of each high and low tide.

An even better machine was put into service in 1910. Tide-Predicting Machine No. 2 is about eleven feet long, two feet wide, and six feet high. The machine has fifteen thousand parts. Not only does it draw a curve of the predicted tide at the place for which it is set, but it also shows on dials the height that the water will be at every minute during the year. At each high and low tide, the machine automatically stops, the operator jots down the hour, minute, and the height of the water, and then he pushes the lever to restart the machine.

Setting the machine to make tidal predictions for a particular place takes about three hours. The machine then gives predictions for an entire year in about seven hours. Ordinarily, the data printed in tide tables is prepared several years ahead of the publication date.

The amazingly accurate tide-predicting machine is also used to compute the speed of currents and the time when they will be strongest. The predictions are published by the government in books called *Tidal Current Tables*.

The facts gathered at tide-recording stations are not only used

U. S. Coast and Geodetic Survey

Tide-Predicting Machine No. 2 is about 11 feet long and has 15,000 parts. In addition to drawing a curve of the predicted tide, it shows on the dials the level of the water at every minute of the day. The machine stops automatically at each high and low tide. The operator fills in the times on a record sheet and then restarts the machine.

for predicting the movement of the sea. They also provide the basis for land measurements. To determine the height of a room, you measure from the floor to the ceiling. But where would you start if you wanted to measure the height of a mountain? The base line is the sea.

Although the level of the sea is continually changing, its average height can be computed from the same tidal data used for making

tide tables. In order to get a really accurate figure, records compiled over a long period of time must be used. Heights of mountains and of all land areas in the United States are measured from
a sea-level figure established in 1929.

The tide records also tell a great deal about changes occurring on
our planet. In most places, the sea level is rising by six or eight
inches every hundred years. Scientists say that this is due to the
slow melting of the Arctic and Antarctic ice. However, at some
places such as Galveston, Texas, the water seems to be rising more
rapidly — about two feet every hundred years. It is believed that
this rise is due both to the increase in the ocean's waters because of
the melting ice and also to the slow sinking of the land. In southeast Alaska, on the other hand, tide gauge records show that the
height of a whole mountain range has increased five feet in fifty
years.

4: Whirlpools and Other Tidal Hazards

The tide stages many spectacular shows when hemmed in by land. In mid-ocean, where there is nothing to interfere with the water's free movement, its rise and fall is gentle. But in coastal areas, where there are numerous shallow spots and many bays, rivers, and straits into which the tide must move, the water not only goes up and down but also flows sideways, frequently with considerable speed. The sideways, or horizontal, movement is called a tidal current. Generally, tidal currents run in one direction for about six hours and in the opposite direction for the next six hours.

In some places, tidal currents create really dangerous conditions. The violent whirlpools formed by tidal currents along the coast of Norway are world famous. They are called maelstroms. *Mael* means to grind; *strom* means stream. One of Edgar Allen Poe's eeriest stories is about a boat that was sucked into a maelstrom and was carried down into the hole, or vortex, at its center. The story is entitled, "Descent into the Maelstrom." Factual descriptions of maelstroms are as awesome as Poe's story.

In a book for navigators published by the United States government, full details are given about the most vicious of all known maelstroms on the northwest coast of Norway. It is called "Saltens'

The violent current in Saltens' Maelstrom on the northwest coast of Norway has thrown ships against the rocky shores.

Maelstrom" because the whirlpool forms near a town named Saltens. Here there is only a narrow channel through which the tidal current must move in and out of a large bay. On the days when the tide rises highest, close to 90,000,000 gallons of water rush through the channel in six hours. The current swirls along at a high speed; at times it is said to reach a velocity of as much as thirty miles an hour.

When the current is flowing at maximum speed, its force is so great, according to the official description, "that the houses near it tremble and whales have been driven back when trying to force a passage through. Even in fine weather the noise of the current may be heard at a considerable distance."

The channel is navigable only during brief periods each day when the current is about to reverse the direction of its flow. Ships wait until that time, called "slack water," to pass through the channel. While waiting, ships must be maneuvered with great care. Ships caught by the current have been hurled against the rocky shore with such tremendous force that they have been smashed to pieces.

Even slower-moving currents create hazards. The *Mayflower,* which brought the Pilgrims to America, came very close to being wrecked by the currents running along the coast of Cape Cod.

At daybreak on November 19, 1620, nearly two months after sailing from Plymouth, England, the *Mayflower's* lookout shouted, "Land ho!" In the bright light of early morning, the land was identified as Cape Cod. This was not the area in which the Pilgrims planned to establish their colony. They intended to settle near the Hudson River. The captain of the *Mayflower,* Christopher Jones, knew that to reach the Pilgrims' objective, he would have to sail to the south. The question was whether to continue or to stop so that the passengers could rest for a while. The Pilgrims decided not to stop.

It was a beautiful morning with a good wind to blow the *Mayflower* toward her destination. Everyone aboard was happy as they sailed along the coast. They felt secure now that they were near land. But late in the afternoon, the *Mayflower* was trapped in a dangerous tide rip. Pollock Rip, as the place is called, is a shallow area with many sandbanks, some only a few feet under water.

The tidal currents of the Cape Cod coast, like most others, flow smoothly as long as they move through uniformly deep channels. But when a tidal stream flows over an uneven sea bottom with many sandbanks or rocks, the water is forced up into vicious waves. They crash in all directions.

The Pilgrims, who had never before seen such a wild-looking sea, were terrified by the roaring breakers. But Captain Jones was more worried about where the current would carry the ship. After the *Mayflower* sailed into the rip, the wind died and the ship drifted helplessly, battered by waves and swept along by the current. Captain Jones, an experienced seaman, was familiar with tide rips. There are many along the English coast where he ordinarily sailed. He knew that the *Mayflower* was in danger. If the current swept her onto a sandbank, the ship might be swamped by the waves. Or the *Mayflower* might be thrown on a bank in such a way that it would be impossible to free her.

A lightship maintained by the United States government marks the dangerous Pollock Rip area on the Cape Cod coast. The Mayflower *was caught in these tide rips and was nearly wrecked.* U. S. Coast Guard Photo

But the *Mayflower* was lucky. The wind started to blow again, and Captain Jones managed to sail out of the tide rip.

The Pilgrims were then faced with another problem, for the new wind was blowing from the south, the direction in which they wished to go. Even with a favoring wind, the *Mayflower* moved slowly; she could not, as can modern sailboats, go to windward. Of course, the Pilgrims might have waited until the wind changed again. But after the frightening experience in the tide rip, they decided to turn back and to head for the nearest safe harbor.

This decision was reached late in the afternoon. Although the distance along Cape Cod to Provincetown harbor, where the *Mayflower* anchored, was less than fifty miles, it took two nights and a day to get there. Adverse currents made the trip slow.

If a boat is traveling in the same general direction as the current, it adds to her speed. A boat moving at five miles an hour in a current flowing at two miles an hour will actually cover seven miles. However, the same boat sailing against the current will make only three miles in an hour. Not only for sailboats but also for ships propelled by engines, tidal currents are of great importance. By traveling with the current, instead of fighting it, speed is increased and less fuel is consumed.

Nowadays, navigators can obtain information about tidal currents so that they can use them advantageously. The data is published in current tables, and for some harbors and sections of the coast, special maps show the direction and speed of the current for each hour of the day.

Gathering the facts about currents is far more difficult than measuring the vertical movement of the tide. The gauges and tide staffs used for measuring how much the water rises and falls are installed near shore and can be read from the land. When currents are tracked, a specially equipped boat must be sent out with a crew of technical experts. Making such a survey takes a long time and is very expensive. A recent survey of the currents in New York harbor took two years to complete. The men who planned the project thought it would take longer.

The work was speeded up by the use of an improved type of radio current meter. In former days, the speed of a current was measured by a rope attached to a pole. The bottom of the pole was weighted with lead so that it would float vertically. The rope was attached near the top of the pole. When an observation was to be made, the pole was lowered into the current, and as it was carried away, the rope was allowed to run out. At the end of exactly one minute, the rope was checked. The speed of the current was computed from measurements marked on the rope.

During a survey, the rate at which a current is flowing is measured at half-hour intervals throughout the day and night. It was slow, tedious work when a pole was used, and the results were not nearly so accurate as are now obtained from radio current meters.

A radio current meter has a propeller, which is turned by the flow of the water and measures its speed. The meter is so constructed that its nose always points into the current. If the current is flowing from the southeast, the meter points to the southeast. The meters are suspended below a float containing a radio transmitter. When the transmitter is activated by a signal from the survey ship, it automatically sends coded messages giving the speed and direction of the current as indicated by the meters.

Since the speed of a current may be different near the surface and farther down, it was decided to record conditions at three levels

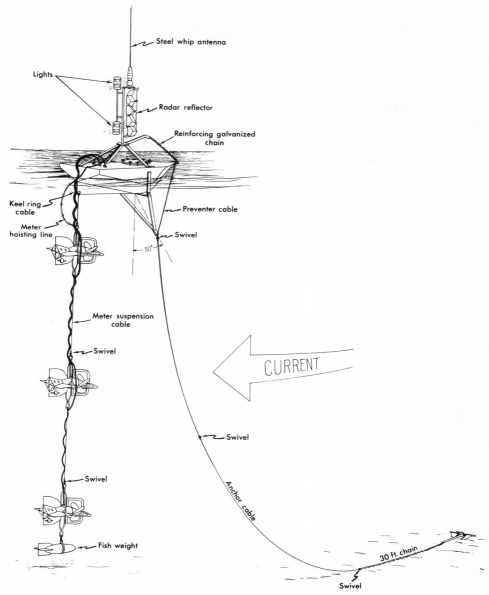

Steel whip antenna

Lights

Radar reflector

Reinforcing galvanized chain

Keel ring cable

Preventer cable

Meter hoisting line

Swivel

50°

Meter suspension cable

Swivel

Swivel

CURRENT

Swivel

Anchor cable

Swivel

Fish weight

30 ft. chain

Swivel

U. S. Coast and Geodetic Survey

Equipment used for measuring the speed and direction of tidal currents. The three meters with propellers record conditions at different depths. The information is transmitted by a broadcasting unit in the float.

during the New York harbor survey. This was easily done by suspending three meters one below the other under each float.

The men on the survey ship anchored four floats in positions from which information was desired. Then, at half-hour intervals, each group of meters was tuned in, and their reports were recorded on tape. When sufficient data had been obtained, the floats were moved to new positions. Thus the tidal streams running through New York harbor were tracked accurately and quickly.

In order to locate a current's channel, dye may be poured into the water. The dye shows the path along which the water is flowing. In a survey made in the harbor of Galveston, Texas, a reddish pink fluorescent dye was used. Airplane pictures were taken of the colored water. In addition, the water was tested with a device somewhat similar to a light meter used by photographers. The color meter showed exactly how much dye remained in the water at varying distances from the place where it had been poured in. The survey indicated that if waste substances were put into the water during a flood tide, they would be carried out of the harbor in twenty-four hours but would not leave the Galveston area for about six days.

Although many currents flow in one direction for a certain number of hours and then turn and flow in the opposite direction, some currents change direction gradually. They move around as if they were following the spokes of a wheel. First, the water flows where one spoke would be, then along the next, and so on until it has completed a full circle. It is important for a navigator to know the angle at which such a rotary current is moving and its speed when his ship is passing through it. Ships cannot be kept correctly on course unless proper allowance is made for the effect of currents.

Timetables can be made of tidal currents because each current moves according to a definite pattern. Although closely related to the vertical up-and-down movement of the sea, tidal currents frequently have an independent time schedule. For example, at a particular place along the coast, the tide may be dead low at twelve o'clock and begin to rise immediately after noon. But the direction in which the current is running may not change until two o'clock or even later.

Therefore, a navigator needs both a tide table to tell him the time of high and low water and a current table to tell him the time when a tidal stream will change its direction and when it will be moving at maximum speed. No tidal current always flows at the same speed. A current that runs first one way and about six hours later in the reverse direction moves slowly immediately after it turns. It works up to maximum velocity about three hours later and then slows down until its direction changes again. Currents are strongest on the days of spring tides when the water rises highest and falls lowest.

Not only in the whirlpool areas on the Norwegian coast but in many other places, ships must wait until the current is almost at a standstill before navigating through it. On one of the main shipping lanes on the Alaskan coast, a violent current runs almost at right angles to the channel. This channel, called Unimak Pass, is between two of the Aleutian Islands. The danger is that the current will push the boat out of the channel and onto the rocks bordering it. The official piloting directions advise that "large vessels should enter only at slack water."

The Aleutian Islands form a barrier between the Bering Sea and the North Pacific Ocean. The tide does not rise and fall at the

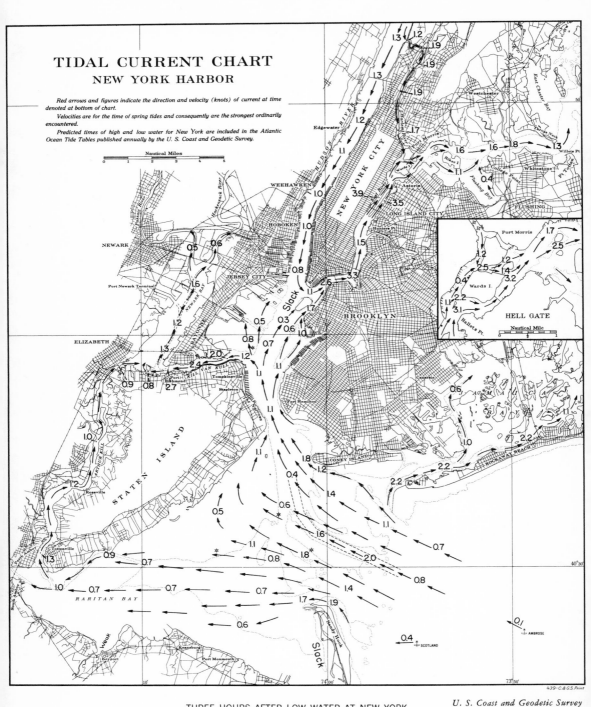

TIDAL CURRENT CHART
NEW YORK HARBOR

Red arrows and figures indicate the direction and velocity (knots) of current at time denoted at bottom of chart.

Velocities are for the time of spring tides and consequently are the strongest ordinarily encountered.

Predicted times of high and low water for New York are included in the Atlantic Ocean Tide Tables published annually by the U. S. Coast and Geodetic Survey.

HELL GATE

THREE HOURS AFTER LOW WATER AT NEW YORK

U. S. Coast and Geodetic Survey

439-C.& G.S.Print

same time in these two bodies of water. While the tide is rising in the Bering Sea, it is falling in the North Pacific, and while it is rising in the Pacific, it is falling in the Bering Sea. The current rushes with great force from the ocean where the water level is higher to the one where it is lower.

There are many straits with such strong currents that boats do not try to battle them. On the Pacific Coast, steamers wait near Vancouver Island until the current slacks off before passing through Seymour Narrows. In the vicinity of New York City, the current in the East River affects the movement of ships. The river, actually a strait connecting New York harbor and Long Island Sound, is used chiefly by barges and other freight carriers. Although ships with powerful engines can make headway against the current at the place called Hellgate, many captains prefer not to waste time and fuel fighting the current. They schedule their passage through the East River when the current will be running their way and will give them a lift.

Captains also use tidal currents when going up or down rivers that empty into the sea. The ocean tide enters the mouth of a river and continues upstream until it is stopped by a dam, a waterfall, or the force of the river's own current flowing toward the sea. The tide moves only a short distance into the Mississippi River both because the tidal movement in the Gulf of Mexico is weak and because the tidal current is stopped by the vast quantity of fresh water pouring down the river. But in the Hudson River the tide

A page from a book of charts showing the speed and direction of the tidal current in New York Harbor for each hour of the day. The chart on page 54 shows conditions three hours after the tide has begun to rise. Note that the current is still ebbing in the Hudson River.

travels one hundred and thirty-one miles upstream before it is stopped by a dam. In the Amazon River in South America, the tide travels nearly five hundred miles up the river.

It is in rivers that the tide produces one of its most awe-inspiring sights. In the Amazon, the Tsientang Kaing in China, the Petit-codiac in Canada, the Seine in France, and in some other rivers, the incoming tide is frequently preceded by a high, forward-moving wall of water. The wave is known as a bore. It occurs only where there is a great tidal rise and fall at the river's mouth and where sandbanks or other obstructions prevent the free movement of the tide into the river. As a result, the water piles up, sometimes reaching a height of as much as twenty-five feet above the level of

Bore rushing up the Petitcodiac River at the head of the Bay of Fundy.

the river ahead of it. The approach of a bore is heard first as a murmur, then as a roar.

The highest known bores sweep up the Tsientang Kaing River. At one observation point, it was estimated that 175,000,000 tons of water rushed by in a minute. In ancient times, the great walls of water rose above the river's embankments and flooded the land. Shortly before one of the most disastrous floods, the Chinese emperor had his most successful general murdered. The man had won so many victories and become so popular that the emperor feared he might try to seize the throne. The general's body was thrown into the Tsientang Kaing River, and when the bore swept up the river and over the land, the emperor thought the general's spirit was seeking revenge.

According to the ancient legend, the emperor tried to make amends by throwing offerings of food and other things into the river. When this proved ineffective, he decided to build a pagoda where the water had caused the worst break in the embankment. The pagoda, or shrine, was supposed to induce the good fungshui, or spirit, to tame the river.

To the emperor's great joy, the flooding did stop. The reason, of course, was that when building the shrine, the embankment was strengthened and its height increased enough to hold in the water.

Even today, superstitious people living near the Tsientang Kaing throw offerings into the river. However, Chinese boatsmen know that the bore moves up the river a certain number of hours after the tide has started to rise in the China Sea, and they keep their boats in protected spots until the wave has passed. Then they launch their boats and ride upstream in the fast current behind the bore.

The bores in the Seine River, on which Paris and other great French cities are located, formerly were a serious menace to ships going up and down the river. By removing the sandbars at the river's mouth, the French government has succeeded in reducing the danger created by the bores. Now, on most days, there is no bore in the Seine. Only when the tide at the entrance to the river rises more than twenty-six feet are bores formed. This happens on an average of twice a month. However, as a result of the dredging, the bores are not as high as they formerly were.

The waterfall at the mouth of the St. John River cascades into the Bay of Fundy when the tide ebbs. The waterfall runs in the reverse direction when the tide rises. The dark portions of the rocks are under water at high tide.

Kosti Ruohomaa from Black Star

Canadian National Railways

For about ten minutes during each rising and falling tide, the water in the bay and the river are at the same level. Then boats can move in or out of the river.

A tidal spectacle of a different kind occurs daily in the St. John River, which empties into the Bay of Fundy. A short distance from the river's mouth, there is a narrow gorge that has, at its seaward end, a reversing waterfall. When the tide rises in the bay, the waterfall faces upstream. When the tide ebbs and the current flows from the river, the water cascades down into the sea. The height of the waterfall may be as much as sixteen feet. Only for about ten minutes during each rising and falling tide is the water in the bay and the river at the same level. During this brief period, there is no waterfall, and boats can safely pass in or out of the river.

5: Neither Sea nor Land

Many creatures of the sea seem to know all about the tide. Small, silvery fish called California grunions know when the highest tides of the month will occur. From March through August, on the nights after a new or full moon, when the tide rises highest, grunions swim out of the sea and swarm on beaches of southern California. The fish arrive at about the turn of the tide, after the water has stopped rising and is about to recede.

Grunions come ashore to bury their eggs in the sand. The female digs herself into the wet sand by wriggling her tail in a drill-like motion. The eggs, released several inches below the surface of the sand, are fertilized by the males that accompany the females. The entire process is completed in about thirty seconds, and the grunions then return to the sea. Until the next spring tide, about two weeks later, the water does not rise high enough to reach the eggs. The top layer of the beach dries, but the sand underneath remains sufficiently damp for the eggs to develop. When the tide again rises unusually high, the waves lap over the buried eggs and wash them out of the sand. The eggs hatch as they are tumbled about in the surf, and the tiny grunions disappear in the sea.

The two-week interval between spring tides is exactly right for

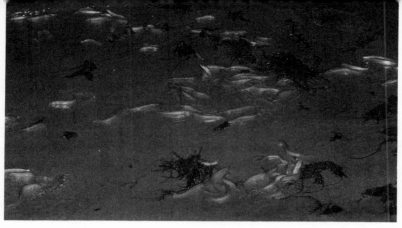

Bucky and Avis from National Audubon Society

Grunions swim out of the sea during spring tides and bury their eggs in the sandy beaches of southern California.

the incubation of the eggs. If the grunions came to the beach at the end of an ordinary high tide, the water of one of the next high tides would flood over the eggs before they had time enough to develop.

Scientists have succeeded in hatching grunion eggs in laboratories by duplicating the natural schedule. The eggs were kept in damp sand for about two weeks, the length of time between one spring tide and the next. Then sea water was poured over the eggs, and they were tossed around in artificially made waves. The eggs hatched, but the young grunions died after about a month. Something was lacking in the laboratory's salt-water tanks that the fish find in the sea.

It is easy to obtain a supply of grunion eggs. All one need do is to check the tide tables to find the dates of the higher than average tides during the months when grunions spawn and then wait on the beach until the fish come in. They come ashore at night, which indicates that grunions have a really uncanny tide sense, for along the southern California coast the water rises higher at night than during the day. If the fish buried their eggs during the day, they would be washed out by the night high tide.

Like grunions, horseshoe crabs leave the sea to bury their eggs on beaches. Horseshoe crabs have probably been doing this for hundreds of millions of years. Fossils of creatures much the same as the horseshoe crabs of today have been found in areas just north of the Alps Mountains. These fossils date back to the era when central Europe was covered by a great sea.

Now horseshoe crabs have disappeared from Europe. They live along the American east coast from Maine to Yucatan and in Asiatic waters from India to Japan. No one knows how horseshoe crabs happened to reach these areas. It is assumed that they were transported by ocean currents, for horseshoe crabs can neither crawl nor swim efficiently. They are awkward creatures with heavy protective shells and sharp tail-like spines. The spine, however, is not a fighting weapon. When a horseshoe crab is thrown over on its back by a wave or anything else, it rights itself by thrusting the sharp end of its spine into the sand and pivoting around on it. Horseshoe crabs spend most of their lives plowing through sand or mud on the sea bottom searching for worms and other soft animals.

Periodically, the animal outgrows its shell and sheds it. At molting time, many discarded shells float in with the tide and are left on beaches. The size of the shells indicates the age of the crabs that shed them. The shell of a two-year-old crab is about six inches in length. When a horseshoe crab is full grown, its shell is approximately two feet long.

Why horseshoe crabs and grunions come ashore to bury their eggs is a scientific mystery. Most creatures of the sea, as well as those that live between the boundaries of high and low tide, discharge their eggs in the water. One of the most important steps in

Fossil of an ancient horseshoe crab. Except that parts of the shell have been worn away, the fossil and a live horseshoe crab look alike.

the evolution of an animal from a marine to a land form is the development of a reproduction system entirely independent of the sea.

The tiny snails called rough periwinkles have passed this step. Their young are born alive on rocks above the reach of the average high tide. The newborn snails are so small that a magnifying glass must be used to see their coiled shells. They are exactly like those of full-grown rough periwinkles.

The snails inhabit rocks flooded only when the tide rises unusually high. The snails can withstand long periods of exposure to the air and sun but not long periods of submersion. Like land animals, rough periwinkles drown if held under water.

Yet these snails still are dependent on the tide. When the high spring tides rise over the rocks or storm waves splash over them, the snails crawl around feeding on microscopic plants. When the rocks are dry, the snails are inactive.

During a laboratory experiment, rough periwinkles gave an amazing demonstration of their tidal instinct. Only at the times when the tide would have reached their native rocks did the snails crawl around. This reaction to the tides' schedule continued for several months.

Some snails moved inland ages ago. Their descendants are frequently found in woods and gardens. Other members of the snail family live near the low-tide mark and are covered by the sea most of the time. The rough periwinkles are at an in-between stage. They cannot yet part from the sea, but they cannot live in it.

Snails with coiled shells are not able to withstand battering by waves as well as can their relatives the limpets, which have cone-shaped shells. Limpets clamp their shells so tightly to the rocks that indentations are made in them.

Limpets clamp their shells so tightly to rocks that grooves are cut in them. Limpets leave their home sites to find food but return to the grooves into which their shells fit perfectly.

Limpets are found on rocks at all levels between the low- and high-tide marks. The level at which a limpet lives has an effect on the height of its shell. Limpets that live near the high-tide mark have taller shells than those living at lower levels. In a taller shell, more water can be stored to sustain the animal during the hours when the tide is out and the rocks are dry. At these times, limpets do not budge. They wait until the tide wets the rocks and then crawl around grazing on minute plants.

But after feeding, limpets return to their home sites. This homing instinct was recorded more than two thousand years ago by the Greek scholar Aristotle. Apparently, limpets feel their way back. When a groove was filed across the path of one animal, it was completely bewildered. After an unsuccessful search for the road home, the limpet settled on an unfamiliar spot and waited until the next tide rose over the rocks. Then the animal continued its search and managed to find its way around the groove. At its home place, there was an indentation in the rock into which the limpet's shell fitted perfectly.

Snails with coiled shells live at all levels of the intertidal zone. Some are covered by water only when the tide rises unusually high.

Limpets are true marine animals and wait until the tide rises over them to shed their eggs into the water. The eggs drift in the sea until the baby limpets hatch and are ready to settle in the tidal zone. In areas with strong tidal currents, limpets may be carried a considerable distance from their starting point before they come ashore. The sea serves as the incubator, and the tide provides the transportation necessary for the establishment of new colonies.

Some shell animals never move from the spot on which they settle. Such animals are entirely dependent on the sea to deliver food to them. The delivery system is most efficient. If it were not, the shell animals called barnacles could not thrive as they do in almost every part of the world.

Barnacles attach themselves with a natural cement to rocks, the foundations of piers, the bottom of swimming rafts, boats, or anything that is under water all or a part of each day. Barnacles have pyramid-shaped shells with two valves at the top that are opened when the sea covers the animal. Through the opening, the animal extends six feathery-like legs to pick food from the water. When exposed to the air, barnacles close their shells, and neither the hot summer sun nor the cold winter wind kills them. But barnacles must be submerged periodically, for unless the sea brings them food, the animals die of starvation.

The most difficult part of the world in which to live is the zone that the tide alternately floods and leaves dry. At high tide, fish swim in from deeper water to feed, and at low tide land birds and animals come to hunt. Yet so perfectly are the inhabitants of the intertidal zone fitted for the conditions with which they must cope that the area is heavily populated. Often animals live one on top of the other. Sometimes plants grow on animals, and sometimes animals fasten themselves to plants.

At low tide, a beach containing billions of animals may seem empty of life. But look closely and you will see that the sand is perforated with small holes, each marking the hiding place of an animal. Some of the burrows may be only inches deep, others several feet.

Barnacles keep their shells closed when exposed to the air. When covered by water, the tops of the shells are opened, and the animals' feathery-like legs are extended to gather food.

Hugh Spencer

A ghost crab at the entrance to its burrow. The animal is almost exactly the same color as the sand.

The burrows of ghost shrimps may extend as much as three feet below the surface. Many animals in addition to the shrimp that dug the tunnel may use it as a refuge. Small crabs may be permanent lodgers, and small fish may swim around in the tunnel while waiting for the return of the tide. When the water rises, the shrimps station themselves at the mouth of their tunnels to gather food from the water.

Ghost crabs, which also dig burrows in the beach, barricade the entrance to their homes with pellets of sand each time the tide rises to prevent the sea from flooding in. The burrow of an adult ghost crab consists of a long, slanting tunnel with a horizontal "room" at the end. The living room is deep enough below the surface of the beach so that the sand around it always remains moist.

On some beaches, ghost crabs must travel a considerable distance from their burrows to the sea. Despite their excellent camouflage — ghost crabs are the color of dry sand — the trip to the water is dangerous. Birds or other animals may pounce upon the crab and devour it. Yet ghost crabs must make the trip to the water's edge to wet their gills.

One observer reports seeing the pale-colored crabs waiting for a wave to foam over them. The crabs did not wade into the water; in fact, they seemed as reluctant as are some humans to getting wet. After a wave splashed over the crabs, the observer noted, they scurried away from the water.

Although ghost crabs remain in their burrows for long periods (it is believed that they hibernate during the winter), they are true marine animals. They return to the sea to discharge their eggs.

When the young crabs come ashore with the tide, they dig burrows close to the water. The second burrow a young ghost crab digs is a little farther up the beach. As the crab grows older, it continues to move away from the sea.

Some crabs became land animals long ago, and ghost crabs may eventually join this group. A hundred thousand years from now, or perhaps a hundred million, the gills of the ghost crab may develop into lungs.

All land animals and plants were originally natives of the sea. First plants got a foothold on shore, and they provided a food supply for the pioneer animals that became land dwellers. Some of these pioneers were encased in shells; others were soft worms.

Scientists have traced, step by step, the evolution of land animals. At one time, fish with bony skeletons settled on land. Their swim bladder, which in the sea was needed to provide buoyancy, became a lung. Lungfish found today in stagnant pools in Africa and Australia are descendants of creatures that left the sea in the dim past. Lungfish can live for months without water when their pools dry up. Other descendants of the first fish to invade the land developed legs.

The move from sea to land was made in slow stages. In the beginning, the creatures stayed in the area where they were exposed

Birds hunting on the beach at low tide.

to the air only at low tide. Later, they lived at the half-tide level, where they were out of water for many hours. Only as the animals became adapted to surviving lengthy periods in the air did they climb up to the farthest place reached by the tide, and then beyond it.

But what explanation is there for the movement away from the sea? Was the intertidal zone overcrowded and the competition for food too great? Whatever the reason, there seems to be no question but that the creature that became man lived for a time in the intertidal zone, which sometimes is part of the sea and at other times part of the land.

6: Harnessing the Power of the Tides

When the French government was considering plans for a new electric generating plant, some engineers recommended that nuclear power be used. But after careful study, the government decided that the best and cheapest way to generate electricity was with power produced by the tides. The decision was a most important one: an investment of a hundred million dollars was at stake. Tidal power was chosen because on the north coast of France the range of the tides is unusually great. The vast movement of the water produces sufficient energy to generate 625,000,000 kilowatt hours of electricity each year.

The idea of using the tide as a source of power is a very ancient one. The Romans built tide mills in England after they conquered and occupied that country. These mills proved so satisfactory that their design was copied for hundreds of years. During the third century, the king of Ireland, Cormac MacAirt, sent to England for a craftsman who could build a tide mill for him according to the Roman design.

The first tide mills were used to grind corn and other grain. Tide mills were erected in many places on the European coast where there was a small, natural inlet that could be enclosed by a dam. Heavy doors, built into the dam, were opened and shut by the tide. When the tide rose, the water pushed against the doors until they opened. The water then flowed in until, at high tide, the level of the water on both sides of the dam was the same. When the tide began to ebb and the water behind the dam tried to run out, the pressure on the doors forced them to close.

The miller now had the water he needed to operate his mill, but he had to wait to use it. Not until the tide had ebbed for several hours could the milling be started. When the water on the seaward side of the dam had fallen considerably below the level of the water in the reservoir, the miller opened a narrow gate in the dam. The water then rushed through a channel, called a sluiceway, to the mill. Since the water was flowing from a higher to a lower level, it developed sufficient force to turn the mill wheel. The miller's only cost for his power supply was the expenditure for building the dam.

As time passed, tidal power was used not only for milling grain but also for other purposes. One English tide mill specialized in making guns. The tide was also used to operate pumps in London's water-supply system. The pumps, installed in the arches of London Bridge, forced water up to a turret from which it flowed down into the city. The business of supplying water was so profitable that Peter Morice, who installed the first pump in 1580, requested permission, four years later, to install a second one. Morice paid two shillings a year, or about forty cents, for the privilege of operating each of the pumps. For more than a hundred years, the water-supply business was owned by Morice's family. They sold it in 1701 for about $150,000.

Many of the colonists who came to America were familiar with

The tide mill at Kennebunkport in which settlers took refuge when attacked by Indians.

tide mills, and a number of them were built in coastal settlements. Some of these old mills are still standing and are now showplaces. The one at Kennebunkport, Maine, which many tourists visit, was the scene of an incident in colonial history about which a number of stories have been written. The mill was built in 1749 and soon afterwards was used as a fortress when the colonists were attacked by Indians. They could not force an entrance into the mill, and the colonists were safe enough within it. But they had no food and were close to starvation by the time a squad of soldiers came to their rescue.

Another tide mill, about ninety miles by road from the one at Kennebunkport, was operated until comparatively recent times. Hodgson's mill at East Boothbay was equipped to saw lumber as well as to grind grain. Many famous sailing ships, including the schooner *Bowdoin,* in which Commander Donald B. MacMillan explored the Arctic, were built of lumber cut in Hodgson's tide mill. When a captain was ready to go to sea, he could obtain from Hodgson the grain needed by the cook to bake bread and cake during the voyage. To boats making their first, or maiden, voyage, the grain ground at Hodgson's mill was presented as a gift.

Many mills were operated with tidal power long after the development of steam engines. Tidal power has the advantage of costing nothing; its disadvantage for milling purposes was that it could be used only when the level of the sea was below that of the reservoir. As a result, the miller had to work at odd hours. If high tide occurred at eleven o'clock at night, the miller might start work at about 1 A.M. and continue through most of the night. The next night he could not begin work until about 2 A.M., since each night the tide rises about an hour later.

When steam-driven equipment was introduced, milling could be scheduled for regular hours. However, the miller had to pay for the fuel he used. Many millers economized by using tidal power whenever possible and steam at other times. This system worked well until milling became a large-scale business. In big mills, steam-driven equipment was used exclusively.

Only comparatively recently did engineers begin to urge that tidal energy be used for generating electricity. If rivers can be used to operate hydroelectric plants, they asked, why not use tidal power for the same purpose?

Back in the 1920's, a plan was developed by Dexter P. Cooper for building a tidal electric generating plant in the state of Maine. The location he chose was in the Bay of Fundy area, which has the world's greatest tides. Cooper spent his summer vacations on an island in the vicinity, and as he watched the water surge in and out, he conceived the idea of "harnessing" the tide, or in other words, of using its power to generate electric current.

Franklin D. Roosevelt, Cooper's neighbor on Campobello Island, was intrigued by the idea, and the two men had many long discussions about it. Cooper hoped to obtain money for his project from private investors but was unable to obtain the millions of dollars needed.

Later, in 1935, during Roosevelt's first term as President of the United States, the construction of a tidal electric generating plant was begun with funds provided by the federal government. It was called the Passamaquoddy (Pas-a-ma-kwod-i) Bay Tidal Project. The bay, named for a local Indian tribe, is partly in the United States and partly in Canadian territory. Therefore, the approval of both countries was needed if the entire bay was to be used for a tidal power project. When Canada failed to approve, the United States decided to use only Cobscook Bay, which is on our side of the international boundary line. However, the name of the project was not changed. It was always referred to as Passamaquoddy, or Quoddy.

The wide mouth of Cobscook Bay is dotted with islands, and the plan was to enclose the bay by constructing a series of dams between the mainland and the islands. Some engineers expressed doubt that the dams could be constructed, for billions of tons of water pour into the bay and out again twice each day as the tide rises and falls. And as always happens when vast quantities of water move through narrow channels, the tidal currents are strong. Yet the first three dams were successfully completed. The cost amounted to about seven million dollars. The United States Army Corps of Engineers, which was in charge of the operation, was prepared to complete the damming of Cobscook Bay, but Congress refused to allot additional money and work was halted. The ships equipped to do deep-water drilling, the cranes, dredges, and all the other special machines, which had been brought to the Maine coast for constructing the dams, were taken away and the workmen departed. A new town built for the men and their families was left vacant.

The dams built between the islands of Cobscook Bay in 1935-37 are used as foundations for automobile highways.

Although from time to time efforts were made to revive the project, it seemed for many years that no use would be made of the Bay of Fundy tides. A new plan for an international tidal power plant, submitted in 1963, drew favorable comments. The plan, prepared by engineers representing the United States and Canada, calls for utilizing both Passamaquoddy and Cobscook Bays. Quoddy would be a huge reservoir measuring one hundred and one square miles; Cobscook would be a reservoir of forty-one square miles. In addition, an auxiliary generating plant would be constructed on the St. John River, the one famous for the reversing falls at its mouth.

The cost of constructing the dams and the generating plants, it was estimated, would be more than a billion dollars. According to

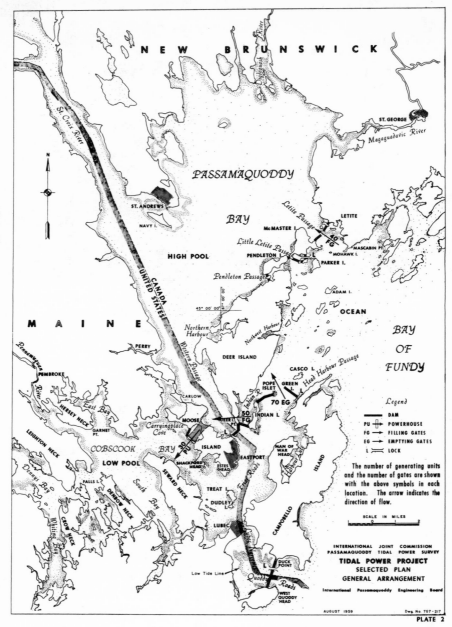

A plan for using both Quoddy and Cobscook Bays to produce electric current.

Engineering News-Record

The world's first tidal hydroelectric power plant is on the north coast of France near the mouth of the Rance River. The location of the dam is shown by the black and white line. St. Malo, a city built in medieval times and now a popular beach resort, is seen in the foreground.

the engineers' report, the plants could generate 1,250,000 kilowatts of current each day. The price at which the electricity could be sold would be lower than that now charged for current generated in plants using oil or coal as fuel.

During the 1963 conference, experts recommended that Congress authorize the Army Corps of Engineers to proceed with the tidal project. Engineers say that about fifteen years will be required for the construction work.

When the Quoddy plan is developed, information will be available from the pioneer tidal hydroelectric plant on the French coast at the Rance River.

Where the Rance empties into the sea, miles of rocky ledges and sandbanks — some said to be composed of quicksand — are exposed at low tide. Signs warning tourists about the tide are posted along parts of the beach used for parking, giving the time when cars must be removed. When the tide rises, many parking areas are covered by deep water, and cars left in them are submerged.

The great tidal range makes the area ideal for a tidal hydroelectric plant. By locating it on the Rance River, the engineers merely needed to build a dam to make a satisfactory reservoir.

Ingeniously designed underwater machines built into the dam can generate electricity at almost all stages of the tide. It is not necessary to wait until the sea level drops below that of the reservoir to generate electricity. The machines regulate the water level. In each of the machines, there is a turbine, generator, and pump. Near the time of high tide, water is pumped from the sea into the reservoir to increase its level. The pumps are then stopped, and the turbines are turned by the force of the water flowing from the reservoir to the sea. Later, when the water level of the reservoir has fallen, the pumps are put to work to speed up the emptying process. After the reservoir's level has been lowered sufficiently, the turbines are turned by water running from the sea into the reservoir.

People use more electricity at certain hours each day than at others. The peak periods of consumption are during the day when factories are in operation and in the evening when many lights are lit in homes. The Rance tidal generating plant was designed to provide current for the peak hours of consumption. The plant is one of many in France's electrical system. All the others burn fuel to produce electricity. By using current produced by the tidal plant during the periods of peak consumption, the demand on other generating plants in the system will be eased.

Before the plant at Rance was completed, Russia announced that it planned to build an experimental tidal hydroelectric plant on the White Sea. Data obtained from the experiment is to be used for

Interior view of one of the underwater machines installed in the Rance River dam.

constructing large tidal power stations along the northwest coast of the Soviet Union.

Although planning and building a tidal station is a lengthy undertaking, it now seems that in the not too distant future there will be a number of tidal hydroelectric plants in various parts of the world.

7: Tides in the Earth and Air

Not only are there tides in the oceans but also in the earth and in the air. According to recent scientific measurements, the earth's crust rises as much as twenty inches twice each day in response to the tide-producing forces of the moon and sun.

For a long time, scientists had surmised that there is such a tidal response in the earth. Every known fact indicated that the gravitational pull of the moon and sun cannot affect the oceans alone. But there was no way of proving the existence of land tides until sufficiently sensitive instruments had been perfected and measuring techniques had been devised.

The particles of which the solid earth is composed do not move as readily as particles of water. Compared with the oceans' tides, the rise and fall of the earth's crust is small.

We can neither see nor feel it, for like passengers aboard a ship, we go up and down with the earth's flood and ebb tides. And not only everyone on earth, but every measuring instrument moves with the earth's crust.

How, under these conditions, can accurate measurements be made? Scientists knew from countless observations and computations the exact strength of the gravitational pull of the moon and sun. These scientific records provided the figures from which it was possible to determine whether the earth's crust moves.

Extremely sensitive instruments were used in making the measurements. One of the instruments is called a horizontal pendulum; another, a gravimeter. The scientists who recorded the land tides made their tests in deep caves to avoid any possibility that temperature changes would affect the accuracy of the measurements. In a cave, the temperature always is about the same. Many other precautions were taken to eliminate errors.

The experiments were conducted by a number of men, each of whom used a different set of instruments and worked in a different cave. All of the tests showed a smaller gravitational pull than had been expected on the basis of the figures generally accepted as a standard. There was only one possible way to account for the lower figures recorded during the cave tests. The solid earth itself must have moved, and the instruments resting on the earth had gone up with it.

After all the calculations had been completed, it was found that there is a movement of the solid earth similar to that of the water in the oceans. Twice a day there is a high tide and twice a day a low tide. The difference between the two may be as great as 20 inches (with an average of 8 to 12 inches).

In addition to the tidal movement of the earth's crust, the flooding and ebbing of the sea causes a downward and then an upward tilting of the land. When the ocean's tide is flooding, many million tons of water move into coastal regions and the weight of the water pushes the land down. When the tide ebbs and the pressure on the land is relieved, its tilt is reversed. The tilting does not stop at the water's edge. It extends far inland.

Pressure is also exerted on the land by the atmosphere. The air above us — the ocean of air, as it is frequently called — exerts an average pressure of 14.7 pounds on each square inch of the surface of the earth. When, due to atmospheric conditions, the weight of the air increases over a given area, the pressure is greater. Cold, dry air is heavier than moist, warm air, and a mass of cold air results in increased pressure on the area over which it is moving. During periods of extremely high pressure, it has been estimated that the earth's crust may be depressed by as much as three inches.

Air pressure is measured by an instrument called a barometer. When the weight of the air (or its pressure) increases, a hand on the barometer's dial moves around in a clockwise direction and points to a higher number. When the weight of the air decreases, the hand moves in a counterclockwise direction to a lower number. Most movements of a barometer indicate a change in weather con-

A weatherman "reading" a barometer. Air pressure is checked and recorded at regular intervals. The instrument in the oblong case is a barograph. It automatically draws a graph of the air pressure.

U. S. Coast Guard

A balloon carries a radiosonde aloft to collect information about the upper air.

ditions. An approaching storm is frequently preceded by a "falling barometer" as weathermen say. What they mean is that the barometer's pointer is moving from a higher to a lower number.

But some barometric changes suggest that there is a vertical movement in the atmosphere every day similar to the rise and fall of the ocean tides. At most places on earth, the barometer tends to move upward until about ten o'clock in the morning. It then falls until about four o'clock in the afternoon, after which it rises again until ten o'clock at night. This rise is followed by another downward movement until four o'clock in the early morning.

The first man to observe this twice-a-day movement in the atmosphere was Pierre Simon de Laplace, who lived in Paris at the end of the eighteenth century. For eight years, Laplace read his barometer four times each day and noted the amount of air pressure it showed. About a hundred years later, the twice-a-day rise of the barometer convinced Sir William Thomson (the man who designed the first tide-predicting machine) that the atmosphere has tides similar to those of the sea. To him, the high barometer at ten in the morning and at ten at night indicated high tides; the lower barometer at 4 A.M. and 4 P.M. indicated low tides.

Unlike the ocean tides, which occur about an hour later each day, the sky tides first observed by Laplace follow the clock. For many years, this puzzled scientists. If the atmospheric tides are caused by the gravitational pull of the moon and sun, why is their tide table so regular? Today many experts believe that the regular atmospheric rise and fall is not due to the effect of gravitation. It is caused, they believe, by the sun's heat.

However, recent investigations show that there are also gravitational tides in the atmosphere. Two scientists who examined the records of hourly barometer readings for sixty-six years reported that there are atmospheric tides caused by the gravitational effect of the moon. These tides are comparatively small. Now the exploration of the upper levels of the atmosphere is producing information about tides sixty to seventy miles above the earth. Some scientists believe that these tides are greater and more important than the atmospheric tides at lower levels that we can measure from the earth.

Usually, we think of tides in connection with the salty water of the seas. But shouldn't the fresh water in lakes also answer the pull of the moon and sun? This question has aroused the curiosity of many scientists, and they have found that fresh water lakes do have tides similar to those of the oceans but much smaller. Measurements of the changes in the water level of three of the Great Lakes at the time of spring tides, the highest of the month, showed rises

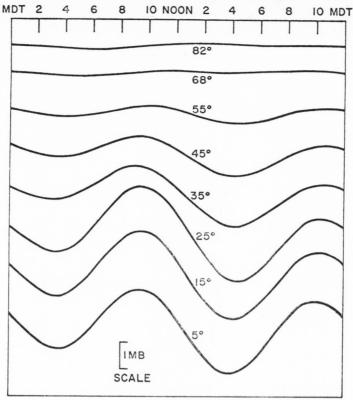

MDT 2 4 6 8 10 NOON 2 4 6 8 10 MDT

82°

68°

55°

45°

35°

25°

15°

5°

1 MB
SCALE

U. S. Department of Commerce, Weather Bureau

Graph showing average daily changes in surface air pressure at various latitudes from near the equator to near the North Pole. The upward and downward movement is similar to the rise and fall of the oceans' tide.

of a little more than three inches. Although lake tides are so small that they are difficult to discern, they are nonetheless fascinating. They provide additional proof that every part of our planet responds to the pull of the moon and the sun.

Of course, gravitational force works two ways. The moon and, to a lesser degree, the sun create tides on our planet. And our planet exerts a strong tidal pull on our satellite, the moon. Since the earth is more massive than the moon, the earth's gravitational force on the moon is far more powerful than the moon's pull on the earth. In fact, it is believed that the earth's gravitational attraction has helped to create vast craters and mountains that can be seen through telescopes. When viewed without any special equipment, the craters, mountains, and plains on the moon sometimes look like a face. They are the "man in the moon."

Lunar experts say that in early times, when the moon was closer to our planet, its thin crust was frequently strained by earth's tidal pull, and the seething lava underneath the crust erupted. First a dome, or hump, was formed by the uprush of lava and gas. Then, after the upheaval ceased, the dome collapsed and sank into the burning hot lava. The resultant melting created a crater. In some areas, it appears that there were many eruptions. Telescopes disclose many small crater areas within larger, older craters.

Sometimes the lava solidified during an eruption, and mountains were formed. And sometimes the lava spread horizontally and made level plains.

View of Clavius taken with a telescopic camera. The crater has towering walls 17,000 feet in height. A number of small craters in the region of Clavius show up sharply.

Mt. Wilson and Palomar Observatories

Photograph of the full moon. The dark areas are lunar "seas."

Latin names are generally used on maps of the moon. A plain is called a "mare," the Latin word for sea. But the moon's seas are waterless. If they were filled with water, the tides in them would rise to tremendous heights because the earth's tide-producing power is so great.

The moon is not a perfect sphere; it is egg-shaped. It has a bulge on the side facing the earth. The bulge, astronomers say, was caused by earth's tidal pull. The earth maintains such a powerful grip on the bulge that, relative to the earth, the moon ceased to rotate. This is the reason that we, on earth, can see only one of the moon's hemispheres. The tidal bulge is toward the center of the side of the moon that it shows to the earth.

The tides on both the earth and the moon have, according to astronomers, affected both our planet and our satellite. The earth, they tell us, at one time was spinning around on its axis much more

rapidly than at the present time. The days were shorter. But the friction caused by the tides slowed down the earth's rotation. The moon was then nearer the earth, and the same force, the experts say, that reduced the rate of the earth's rotation pushed the moon farther from us.

Now the moon is far enough from the earth so that the effect of the tides on the earth's rotation is exceedingly small. The earth is slowing down about one minute in six million years. Even this is enough to cause the moon to move farther out into space.

Some scientists have suggested that about the year 50,000,000,-000 the moon will be such a great distance from the earth that its pull will be too weak to cause tides. But the sun's tide-producing power will not be affected, and the solar tides will continue to slow down our planet's rotation. The eventual result will be that the moon will again move closer to the earth. There is a possibility, according to astronomers, that the moon may come so close to the earth that it will be shattered by the earth's gravitational force. The three rings surrounding the planet Saturn may be composed of fragments of a moon that once orbited around Saturn. The date of any such catastrophe to the earth's moon is so many billions of years ahead that no one can imagine what our planet will then be like or whether men will still inhabit it.

Certainly, no one can predict how the tides would act if, instead of a single moon, there were many moon fragments. The idea has the thrill of fiction. But from a practical point of view, there is no purpose in thinking about it, for the tides as we know them will continue for ages to come. The moon and sun will make the seas and the land rise and fall. Moonlight will brighten our planet at night, and the sun's rays will warm it during the day.

About the time the bright path of the full moon is seen on the water, the tide rises unusually high.

Morris Rosenfeld

Index

Page numbers in italics refer to illustrations